S-CLASS DESTROYER PIET HEIN
EX HMS SERAPIS

FLEET DESTROYERS

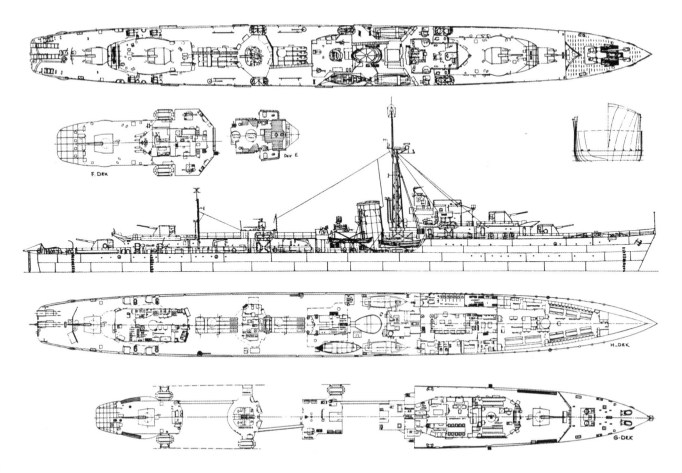

S-Class Destroyer Piet Hein.

Destroyers are small warships (defined in the London Treaty of 1930 as being no more than 1,850 tons), and armed with light weapons (guns of calibre no more than 5.1 inches (130 mm)).

In the Second World War a Royal Navy ship also had to be fitted with torpedo tubes to be classed as a destroyer. Usually, these ships were equipped for anti-submarine work, although some were equipped for minelaying operations.

Before the Second World War new British destroyers were generally designated as fleet destroyers, for work in support of the main fleet, which included cruisers and capital ships. Experience during the war led to older destroyers often being refitted and re-designated as escort destroyers (principally because older destroyers lacked the speed of modern warships) and used for less glamorous tasks such as convoy escort. The need for new specialist escort destroyers was recognised, however, and these were also built during the war.

The fleet destroyers, equipped for anti-submarine work. Each class was often fitted with improved anti-aircraft armament compared to previous British destroyers. Nevertheless, (as with all small ships) they were extremely vulnerable to air attack.

With the need for more destroyers quickly, the inability to build a prototype for testing, the difficulty of future modifications and the potential consequences of failure it can perhaps be understood why many subsequent fleet destroyers adopted a slightly simplified version of the same hull form. A new class incorporated hard-won war experience and was based on previous destroyers. With the same power plant, basic hull form, identical speed, and similar main weapons (generally, with slightly reduced guns to speed up construction and reduce costs). Anti-submarine capability along with anti-aircraft armament increased displacement and resulted in less freeboard).

Tribal class destroyer.

PIET HEIN

October 1952: Piet Hein *passing a US Navy vessel in Shimonoseki Strait.*

INTRODUCTION

The homeland was destitute. Not only had the fleet been greatly affected in quantity and quality by WWII, but the country had also suffered particularly badly. Cities and villages, factories and shipyards, harbours, canals, bridges, and locks were mostly destroyed or damaged.

The reconstruction of the navy started with surplus warships from the war stocks of foreign navies. In September 1945, the Netherlands agreed to purchase four British destroyers to help to re-equip the Royal Netherlands Navy, the three S-class destroyers *Serapis*, *Scorpion* and *Scourge* and the Q-class destroyer *Quilliam*. *Serapis* was commissioned into service in the Netherlands on 5 October 1945, with the new name *Piet Hein*.

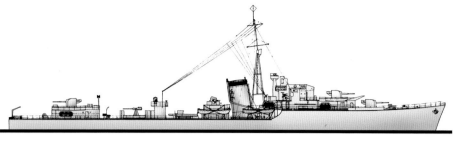

Left: HMS Jupiter *in late 1940. Eight ships formed the J-Class Despite the heavy losses it was generally accepted that the design was sound, and formed the base for the 'emergency' designs that followed.*

After WWII broke out the 'War Emergency' programme required a standard destroyer design to build in quantity, and this design was modified for the purpose. The main change was to go back to the simplicity of four single gun mounts, but the design was also modified to make the ships slightly smaller - and then the next groups had to be slightly enlarged - so some of the advantages of stabilisation were discarded.

HMS Partridge *one of the early War Emergency Programme destroyers in late 1941.The P-Class did incorporate war experience into their design from the start and had much improved air defence capabilities.*

British Admiralty ordered the eight destroyers of the S class on 9 January 1941 as the 5th Emergency Flotilla. Generally similar to previous Q- and R-Class destroyers but with a Tribal-Class bow to make them less wet forward. Intended for general duties, including use as anti-sub-marine escorts, they had to be suitable for mass-production. They were based on the hull and machinery of the pre-war J-class destroyers, but with a lighter armament (effectively whatever armament was available) in order to speed up production.

The S-Class destroyers consisted of 8 ships. The 4.7-inch guns were in improved mountings with 55° elevation and better ammunition supply, and most of the ships had a triaxial twin Bofors replacing the quadruple pompom. *Savage* had an experimental twin 4.5" gun turret in 'A' position.

S-Class Destroyers						
Name	Pennant	Builder	Laid down	Launched	Commissioned	Remarks
Savage	G 20	Hawthorn Leslie	7 Dec. 1941	24 Sep. 1941	8 June 1943	For the majority of her wartime career, *Savage* supported Arctic convoys. Scrapped in 1962
Saumarez	G 12	Hawthorn Leslie	1941	20 Nov. 1942	1 July 1943	Flotilla leader, heavily damaged by a mine on 22 October 1946. Written off as a constructive total loss and sold on 8 September 1950 for scrapping Broken up in Charlestown, Fife in October 1950.
Shark	G 03	Scott's & Co.	5 Nov. 1941	1 June 1943	11 Mar. 1944	Became Norwegian *Svenner*. The ship was hit by two torpedoes fired from one of two German torpedo boats, of the 5th Torpedo Boat Flotilla operating out of Le Havre.
Serapis	G 94	Scott's & Co.	14 Aug. 1941	25 Mar. 1943	23 Dec. 1943	Became Netherlands *Piet Hein*
Success	G 26	J.S. White	25 Feb. 1942	3 Apr. 1943	26 Aug. 1943	Became Norwegian *Stord*. She played an important role in the Battle of the North Cape sinking of the German battleship *Scharnhorst*. Officially purchased from the UK government in 1946 and scrapped in Belgium in 1959.
Swift	G 46	J.S. White	12 June 1942	15 June 1943	12 Dec. 1943	*Swift* participated in the Normandy landings providing fire support and was sunk off Sword Beach by a mine on 24 June 1944.
Scorpion	G 72	Cammell Laird	19 June 1941	26 Aug. 1942	11 May 1943	Ex: *Sentinel*. In October 1945, Netherlands *Kortenaer* where she saw action in the Korean War and the West New Guinea dispute. Scrapped in 1963.
Scourge	G 01	Cammell Laird	26 June 1941	8 Dec. 1942	14 July 1943	At sea during the Battle of North Cape in 1943, escorting the Russia-bound Arctic convoy JW 55B. Took no part in the fighting. Became Netherlands *Evertsen* where she saw action in the Korean War and the West New Guinea dispute.

In March 1944 Scorpion was assigned to the "Ocean Escort" force for Convoy JW 58, one of the largest Arctic convoys of the war. All ships arrived safely, and Scorpion returned with Convoy RA 58.

Centre: Scorpion *covered* Duke of York *as she returned west to refuel in Akureyri in Iceland on 21 December 1943.*

Scorpion *at 28 knots moving to the rear of convoy R.A.64 in March 1945 to take over the rear guard as an air attack was threatened. The forward gun barrels, shields, breakwater, and deck are glistening with ice.*

Model Plans

Plans are available at:
1- Netherlands Ministry of Defence: www.defensie.nl/onderwerpen/modelbouwtekeningen
2- NVM (Neth. Modellers Association): www.modelbouwtekeningen.nl

Sectional drawing of Serapis *after reconstruction to fast frigate.*

Technical Data Serapis / Piet Hein			
Displacement:	1710 BRT (standard) / 2505 (deep load)		
Length:	362 ft 9 in (110.56 m) oa		
Beam:	35 ft 8 in (10.87 m)		
Draught:	14 ft 2 in (4.32 m)		
Machinery:	Parsons geared turbines, 2 Admirality 3-drums boilers, 2 shafts, 40,000 shp		
Speed:	36.75 knots		
Range:	2,800 miles at 20 knots		
Complement:	180 men*		
Armament:	**As designed:** 4x 4.5" guns (4x 1) 2x 40mm AA (1x 2) 6x 20mm AA (2x 2, 2x 1) 8x 21" torpedo tubes	**Royal Neth. Navy 1946:** 4x 4.5" guns (4x 1) 2x 40mm AA (6x 1) 8x 20 mm AA 8x 21" torpedo tubes	**Royal Neth. Navy 1947:** 4x 4.5" guns (4x 1) 4x 40mm AA (6x 1) 6x 20 mm AA 8x 21" torpedo tubes 2x depth charge throwers 2x depth charge rails

* Netherlands Navy: 233 men

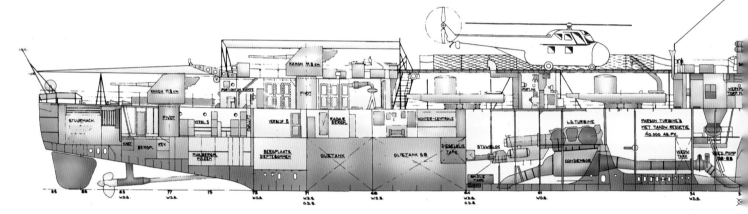

This was a turret in an enclosed 80° twin mounting designed to be fitted in the later 'Battle'-class destroyers. She had no single 4.5" gun fitted in 'B' position, So like her sisters she had also 4x 4.5" guns fitted. Initially *Scorpion* had Pompom instead of Bofors, whilst *Savage* and *Swift* had only 20 mm. The light AA armament at the end of the war was usually increased by one to five single Bofors and up to twelve 20 mm. And the depth charge outfit was 70 or 130. Delays in the delivery of guns, director control towers, fuze keeping clocks and searchlights seriously affected the completion dates of the ships. The ships of the 5th Emergency Flotilla were fitted for Artic service the boat complement was reduced by one 27 ft whaler and davits, and the

After commissioning and workup, Serapis *joined the 23rd Destroyer Flotilla of the Home Fleet based at Scapa Flow.*

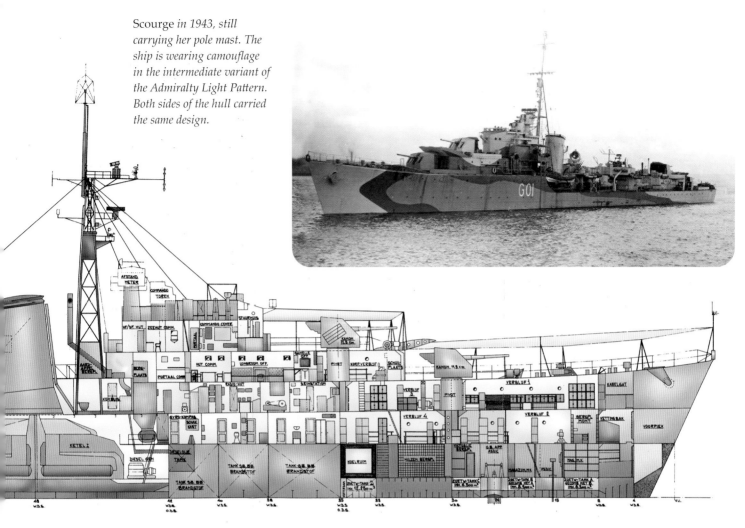

Scourge *in 1943, still carrying her pole mast. The ship is wearing camouflage in the intermediate variant of the Admiralty Light Pattern. Both sides of the hull carried the same design.*

motor cutter was resited at the break of the fo'c's'le owing to damage sustained in its original position when sea came on board.

Serapis was laid down at Scotts shipyard in Greenock on 14 August 1941 and was launched on 25 March 1943. She was completed on 23 December 1943, and assigned the pennant number G94.

Lattice mast

The ever-increasing additions of radar and other antennae had brought serious vibrations in tripod masts which were weakly cross stayed with a triangle of horizontal stays halfway between the base and top. The load at a height of some 12 metres made 'the mast sway like a Hawaiian dancer' necessitating extra stays.
The short-braced lattice masts proved satisfactory, but were not popular with many C.O.s who thought they prevented sighting of aircraft attacking from astern. The mast weighted about 4.5 tons, rigging

and blocks a further 1.5 tons, a slight saving of weight resulted from placing the mast ladders on the fore side instead of after, enabling signalmen to go aloft from the top of the flag locker instead of going down to the upper deck.

By 1945 about 71¼ tons had been added since commissioning, necessitating at least 20 tons of ballast as soon as possible. In return some C.O.s ordered some minor weight savings to be carried out.

Camouflage

In WWII there was widespread use of camouflage on seagoing ships. Although eye catching it was generally felt to be one of the minor aspects of the many wartime efforts and its effectiveness was often impossible to quantify. The philosophy behind the multicolour disruptive schemes was that in any one condition of light and sea state at least one of the tones would blend in with the background while those

which did not would be sufficiently fragmented to either impede or delay recognition and judgement of bearing by breaking-up the ship's structure and interfering with the visual perception of key structural features.
Initially the ships had the Admiralty Light Disruptive Pattern. A scheme that was introduced in 1942 and consisted of blues and greys applied in multiple combinations on hulls. (Decks were painted G20). The tops of masts and crow's nests were painted white. This scheme was designed for northern regions, where hazy and overcast conditions dominated.
By 1944 the camouflage schemes were replaced by a standard simpler scheme (see bottom of Page 5). This scheme was designed to replace most of the existing schemes and was used untill the end of the war.

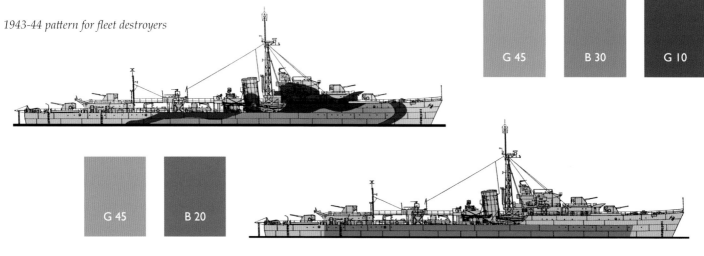

1943-44 pattern for fleet destroyers

G 45   B 30   G 10

G 45   B 20

1944-45 standard scheme decks G 10

G 45   G 10

1945 Admiralty alternative style

The working area of the bridge deck of **Saumarez** *seems somewhat crowded. Note the glass windscreen on the fore bridge; the gyro compass binnacle; and the secondary magnetic compass flanked by voice pipes to the wheelhouse and captain's sea cabin.*

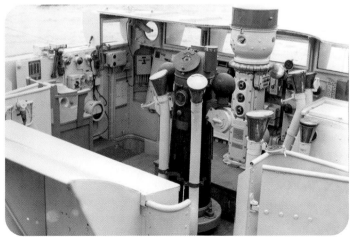

Centre and bottom:
Some details of the bridge of Piet Hein *in the 1950s*

PIET HEIN

Hein was born in Delfshaven (now part of Rotterdam). He became a sailor while he was still a teenager. In his twenties, he was captured by the Spanish, and served as a galley slave for about four years when he was traded for Spanish prisoners. Between 1603 and 1607, he was again held captive by the Spanish, when captured near Cuba.

In 1607, he joined the Dutch East India Company and left for Asia, returning with the rank of captain (of the Hollandia) five years later. In 1618, when he was captain of the Neptunus, both he and his ship were pressed into service by the Republic of Venice. In 1621, he left his vessel behind and travelled overland to the Netherlands.

When the West India Company (WIC) was founded, he was a director of the chamber of Rotterdam. In 1623 he was appointed vice admiral by the WIC. In the following year he captured San Salvador in All Saints Bay from the Portuguese. As admiral and captain-general of the WIC (since March 1626), he captured many Portuguese and Spanish ships.

In 1628, during the Eighty Year's Dutch liberation war from Spain, Admiral Hein, with Witte de With as his flag captain, sailed out to capture a Spanish treasure fleet loaded with silver from the Spanish American colonies and the Philippines. Sixteen Spanish ships were intercepted and captured: one galleon was taken after a surprise encounter during the night; nine smaller merchants were talked into a surrender; two fleeing small ships were taken at sea; and four fleeing galleons were trapped on the Cuban coast in the Bay of Matanzas.

Hein was the first and the last to capture a large part of a Spanish treasure fleet which transported huge amounts of gold and silver from Spanish America to Spain. The amount of silver taken was so big that it resulted in the rise of the price of silver worldwide and the near bankruptcy of Spain.

The capture of the treasure fleet was the Dutch West India Company's greatest victory in the Caribbean. It enabled the Dutch, at war with Spain, to fund their army for eight months (and as a direct consequence, allowed it to capture the fortress 's-Hertogenbosch), and the shareholders enjoyed a cash dividend of 50% for that year. The financial loss strategically weakened their Spanish enemy. Hein returned to the Netherlands in 1629, where he was hailed as a hero. That same year, on June 18, 1629, Piet Heyn was killed in a battle with Flemish privateers at Dungeness. He is buried in the Oude Kerk in Delft.

DEVELOPMENT

When the requirements for the 5th Emergency Flotilla were being considered in the autumn of 1940, the chilling effectiveness of dive-bomber attack had been well and truly appreciated and in consequence there was even more pressure for destroyer main armament to be dual purpose. To this end, 55° elevation was demanded, but against some opposition which questioned the effectiveness of such an intermediate solution and instead proposed twin 4-inch guns as in the Hunt-Class.

navies at the time. The gun had been put into production for the Royal Navy and manufacturing problems had been sufficiently overcome to be able to plan the equipment of the 5th Emergency Flotilla with it, replacing the long-serving 2pdr pom-pom. To give a better field of fire for it, the searchlight was moved forward to the funnel and the Bofors replaced it

between the tubes. Four single Oerlikons were to complete the armament.

Because the J-class hull form had given rise to spray and wetness forward, Tribal design bows were adopted. Otherwise, the only visually distinctive features of the class were the absence of a quadruple 2pdr and the fitting of new-pattern shields to the 4.7-inch guns. The machinery and other arrangements were generally as in the J class.

J-class destroyer HMS Javelin

Emergency destroyers

The British built 112 destroyers in World War II. These were based on the hull and machinery of the earlier J-, K- and N-class destroyers of the 1930s. Each batch produced consisted of 8 destroyers. Because of supply problems and persistent failure to develop a suitable dual-purpose weapon for destroyers, the ships were fitted with whatever armament was available. Advances in radar and weaponry were incorporated as they came available. As a result, they were a relatively heterogeneous class incorporating many wartime advances, but ultimately based on a hull that was too small and with an armament too light to be true first-rate vessels equivalent of their contemporaries.

The argument eventually resolved itself into the 55° mounting but not before consideration was given to purchasing 5-inch DP mountings from the USA. For the close-range AA outfit, discussions ranged around various combinations of quadruple 2pdr, numerous Oerlikons and the new radar-directed Hazemeyer 40mm Bofors Mk IV gun. The latter had been sent to sea before the war by the Dutch Navy and was far in advance of anything developed by the Royal, US or Japanese

Programme			
Class	Flotilla	order date	Remark
O	1st Emergency	3 Sep. 1939	3 ships of O and all P class were fitted with 4-inch guns with a new design of tall gun shield. As a result, they carried only the Rangefinder-Director Mark II(W) for fire control.
P	2nd Emergency	2 Oct. 1939	
Q	3rd Emergency	Mar. 1940	From the Q and R class onwards a transom stern was incorporated.
R	4th Emergency	Apr. 1940	From R class onwards the officer's accommodation was forwards, instead of aft as was traditional RN practice
S	5th Emergency	9 Jan. 1941	Altered the position of the searchlight between the torpedo tubes with the medium anti-aircraft position abaft the funnel. Improving the anti-aircraft gun arcs of fire in the forward field.
T	6th Emergency	Mar. 1941	Lattice foremast, to support the ever-increasing weight of masthead electronics.
U	7th Emergency	12 Jun. 1941	
V	8th Emergency	1 Sep. 1941	
W	9th Emergency	3 Dec. 1941	Dual-purpose Director Mark III(W) introduced, replacing the low-angle Destroyer DCT and High-Angle Rangefinder-Director Mark II(W) in use since the Q and R class.
Z	10th Emergency	12 Feb. 1942	New dual-purpose Director Mk I Type K and the 4.5-inch gun in single mountings CP Mark V as trialled in *Savage*.
Ca	11th Emergency	16 Feb. 1942	
Ch	12th Emergency	24 Jul. 1942	The dual-purpose Director Mk VI introduced with full remote-power control (RPC) for gun laying. One set of torpedo tubes removed to counter increased topweight.
Co	13th Emergency	24 Jul. 1942	
Cr	14th Emergency	12 Sept. 1942	

HMS Savage differed from the rest of the class in being fitted with a new 4.5-inch (114 mm) gun, with a twin mounting forward and two single Mk IV guns aft. The twin mount was taken from spares for the aircraft carrier Illustrious.
In December 1943, the destroyer took part in the Battle of the North Cape which saw the destruction of the German battleship Scharnhorst. After the war, Savage was refitted as a gunnery training ship. The ship was decommissioned and, on 11 April 1962, sold to be broken up.

By the time that *Scorpion* completed, there was a power-operated twin 20mm gun available, which was fitted in lieu of the four single guns, two being staggered abaft the funnel. However, the 40mm twin was not immediately available and she carried a fifth twin 20mm in lieu. She and *Saumarez* also had a lattice mast aft for an HF/DF aerial. *Scorpion*, *Saumarez* and *Savage* completed without lattice masts. Modifications: *Scorpion* had her fifth twin 20mm replaced by a quadruple 2pdr and never in fact received the Mk IV Bofors. Lattice masts were fitted to all except

Savage, with radar 272 and 291 at the masthead. In 1945 radar 276 replaced the 272 set and in that year, *Saumarez* received four single 40mm Bofors on the former searchlight platform consequent upon her deployment to the Indian Ocean.
All served with the Home Fleet as the 23rd Destroyer Flotilla in northern and Arctic waters. In December 1943, *Saumarez*, *Savage*, *Scorpion* and the Norwegian *Stord*

were screening *Duke of York* and *Jamaica*, which had been sailed as distant cover for Convoy JW55B (of which *Scourge* formed part of the escort) in the expectation that *Scharnhorst* would attempt to attack it. The expectation turned to reality when the German battlecruiser sailed from Altenfjord on 25 December, screened by the 4th Destroyer Flotilla comprised of Z29, Z30, Z33, Z34 and Z38.

Scourge (later: Evertsen) was delayed until 8th December 1942 due to serious design problems and the late delivery of armament and fire-control equipment. She was the 11th RN ship to carry this name.

The following day, after engaging British cruisers screening the convoy, *Scharnhorst* retired but was eventually brought to action during the afternoon by *Duke of York* and the S-class destroyers. In the engagement which followed, all the destroyers got into action with gunfire and fired a total of 28 torpedoes claiming four hits. *Scharnhorst* was sunk by a combination of heavy guns of the battleship and torpedo hits from the destroyers, but she badly hit *Saumarez* in the director and rangefinder and near misses caused many casualties and damage from shell-splinters.

During 1944, the flotilla took part in carrier sweeps against the Norwegian coast and in March, while escorting convoy RA57, *Swift* narrowly avoided a torpedo from *U-739*. The flotilla's next task was as part of the massive naval presence at the Invasion of Normandy in June 1944. Here they took part in anti-S-boat sweeps and defended the Invasion area from attacks by German torpedoboats. However, the Norwegian *Svenner* was lost, and *Swift* was mined off the beachhead. After Normandy, the flotilla returned to Arctic duties involving a mixture of Russian convoys and offensive carrier sorties against the Norwegian coast. This task extended until the end of the war in Europe, except for *Saumarez* which went to the Eastern Fleet early in 1945.

Saumarez took part in the carrier raids in the Indian Ocean including the Sabang raid. Now leader of the 26th Destroyer Flotilla, she once again had the opportunity to act in the classic destroyer role when, on 16 May 1945, her flotilla (*Verulam*,

Saumarez. On 22nd October 1946 whilst on passage in the Corfu Channel with Volage *both ships struck mines.* Saumarez *sustained major damage and was towed clear of the area by the damaged* Volage. *There were 43 fatal casualties.*

Fore castle of Piet Hein.

Scourge *while making speed.*

Virago, *Vigilant* and *Venus*) torpedoed and sank the cruiser *Haguro*. *Saumarez*, however, received several shell-hits, one in her boiler room.

Only *Savage* saw significant post-war service under the White Ensign. *Saumarez*, serving in the Mediterranean in 1946, was one of the destroyers involved in the Corfu incident (early episode of the Cold War) when she hit mines laid in international waters off the coast of Albania that October. She was badly damaged when she hit a mine and suffered numerous casualties. Although brought home she was never repaired. Because of the incidents, Britain broke off talks with Albania aimed at establishing diplomatic relations between the two countries. Diplomatic relations were only restored in 1991.

Piet Hein: *note the British colours and pennant number.*

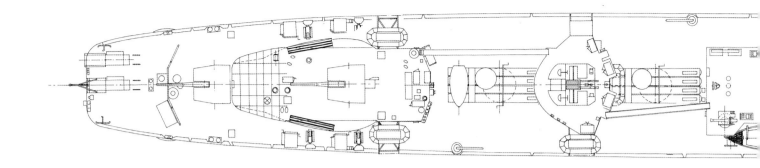

SERAPIS BECOMES PIET HEIN

In October 1945 Serapis was transferred to the Royal Netherlands Navy for service in the Far East, The twin Oerlikons being replaced by single Bofors.

Melbourne 1946.
(Collection: W.H. Moojen)

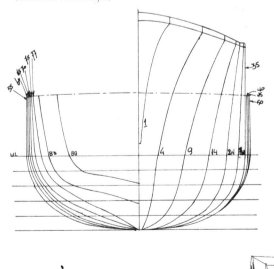

Pennant		
Serapis	G 94	Dec 1943
Piet Hein	J 4	Jun 1946
Piet Hein	JT 4	Jan 1947
Piet Hein	D 805	Oct 1950
Piet Hein	F 804	Oct 1957

Post war RN destroyers

In World War II the Royal Netherlands Navy lost all pre-war destroyers. Replacements were constructed during the war in British shipyards. These destroyers were retained for the post-war navy:

Name	ex HMS	Pennant	Class
Banckert	Quilliam	G 09	Q
Evertsen	Scourge	G 01	S
Kortenaer	Scorpion	G 72	S
Marnix	Garland	H 37	G
Piet Hein	Serapis	G 94	S
Tjerk Hiddes	Nonpareil	G 16	N
Van Galen	Noble	G 84	N

Named *Piet Hein* (Piet Heyn)

1 1774 - 1799

The ship of the line *Piet Hein* was built in 1774 at 's Lands Werf in Amsterdam by John May and had 54 guns. The ship was also called *Admiral Piet Hein*. On August 5, 1781, the ship took part in the Battle of Dogger Bank against the British fleet. On August 13, she served in the North Sea. Shortly afterwards, in 1797, the *Piet Hein* became the flagship of Vice Admiral Adriaan Braak, who sailed to the West Indies with a squadron of 8 ships. In 1798 the ship was rejected in Suriname and sold for scrap in 1799.

2 1799 - 1800

To replace the rejected ship of the line, a Portuguese pink called the *Princess Royal* was purchased in Suriname in 1799, and named *Piet Hein*. She sailed to Bergen in Norway during the course of 1799 with a squadron. *Piet Hein* was sold there on June 16, 1800.

3 1799

According to a story from the navy agent dated October 30, 1799, a privateer ship, the ten-piece brig *Piet Hein*, was purchased. The ship was renamed soon thereafter.

4 1803 - 1809

The schooner *Piet Hein* was built in Rotterdam in 1803 and carried 7 guns. In 1805, the ship took part in the Vlissingen expedition to support the French invasion fleet that was preparing to land on the British coast, which incidentally never happened. In 1808 the ship was assigned to the squadron on the Wadden Sea and on May 21, 1809, taken at the Vlie by sloops of the British 74-gun ship of the line *Princess Caroline*. The commanding officer was seriously injured while defending his *Piet Hein*.

5 1813 - 1819

The 74-gun ship of the line *Piet Hein* was launched in May 1813 in Rotterdam. This ship replaced the *Admiral Piet Hein* of 80 guns, which had already been laid up in March 1806, but which had been fired on the slipway and was subsequently scrapped in 1811. The *Piet Hein* served exclusively in the Netherlands and was sold for scrap in 1819 in Vlissingen.

6 1896 - 1913

Iron cruiser, commissioned on January 3, 1896. In Turkish waters in 1896-1897. Departed for the East Indies on January 7, 1899. Joined the division and left for China on October 9, 1901 to protect Dutch interests there. In 1905 *Piet Hein* made several journeys to the Atlantic, Mediterranean, Norwegian fjords and the Baltic Sea, including a visit to Iceland. On April 22, 1913 the ship was decommissioned and used as a target ship for torpedo boats. Stricken in 1914.

7 1928 - 1942

Admiral-class destroyer commissioned on January 25, 1928. On December 15, 1928, sailed to the Dutch East Indies. Returned in 1934. On November 29, 1934, sailed with *Evertsen* to the Dutch East Indies. Damaged on 13 October 1938 during exercises in the Sunda Strait after a collision with the cruiser *Java*. Since May 10, 1940, on escort duties. In action at Kangean on 4 February 1942. On 15 February 1942, in action at the Gaspar Strait. Sunk on 19 February 1942 during combat in Badoeng Strait by gunfire and torpedoes from the Japanese destroyers *Oshio*, *Michishio* and *Asashio* at 08.40 S.B. and 115.20 O.L.

8 1945 - 1962

S-class Destroyer and subject of this book.

9 1981 - 1998

Kortenaar-class, S-frigate built at KM de Schelde in Vlissingen. In 1998, the vessel was decommissioned and was sold to the United Arab Emirates Navy where she was renamed *Al Emirat*. In 2009, rebuilt into a yacht named *Yas*.

GUNS

The 4.7" (12 cm) QF Mark IX was used on most British destroyers built in the 1930s and 1940s. The S-class, introduced the CP (central pivot) single Mark XXII mounting for the QF Mark XII 4.7 in gun. This new mounting had a shield with a sharply raked front, to allow increased elevation (to 55 degrees), contrasting noticeably with the vertical front of the previous CP Mark XVIII, and easily differentiated the S-class onwards from their immediate predecessors. (*Savage* was the exception in this respect, being fitted with 4.5-inch calibre guns; a twin mounting forward and two singles aft.). These ships used the Fuze Keeping Clock HA Fire Control Computer. It was 3 - 4 times the weight of the previous class's twin gun. On a destroyer this matters a lot.

By choosing "lightweight" open mountings, the designers paid a weight penalty in having to add these shield structures. Bad weather conditions such as those experienced by many of these ships on Arctic convoy duty during World War II greatly reduced the rate of fire.

One of the more noticeable features of destroyers with these guns was the large "spray shield" forward of "B" gun

platform and aft of "X" gun platform. Although these structures did reduce spray on the upper decks, their true purpose was to reduce the problem of blast from the upper guns disrupting the crews of the guns on the lower decks.

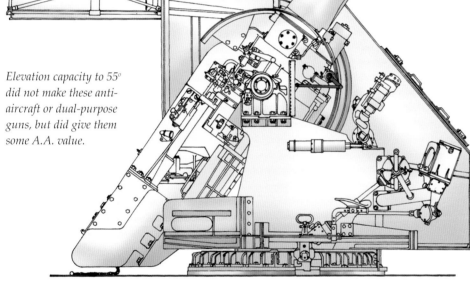

Elevation capacity to 55° did not make these anti-aircraft or dual-purpose guns, but did give them some A.A. value.

Normally, the crew comprised seven men:
1- Gun layer (controls the gun in the vertical plane, and fires the gun when in local control)
2- Breech worker/gun captain
3- Tray worker (operates the gun loading tray and automatic rammer)
4- Trainer (controls the gun in the horizontal plane)
5- Projectile Supply
6- Cartridge Supply
7 Sight Setter (Responsible for communication between transmitting station and gun captain, that gunsights are properly set, and when necessary, time fuzes are accurately set by the fuze setting devices).

In addition, there were extra crewmen assigned to resupplying the ready-use lockers with shells and cartridges sent up from lower quarters, though they were not actually numbered members of the gun crew.

"A" gun while in Korea War. Note the painted national ensign on top.

Technical Data	
	4.7"/45 (12 cm)
Designation	Mark XII
Date in service	1930
Calibre	4.724 inches (120 mm)
Rate of fire	10 - 12 rounds per minute
Max range (40°)	16,970 yards (15,520 m)
Shell weight	HE: 50 lbs. (22.68 kg)
Train rate	Manually operated
Elevation	-10 (?) / +60 degrees
Train	About +150 / -150 degrees "X" position: 360 degrees
Total weight	Mark XII: 3.245 tons (3,297 kg)

Hazemeyer gun mount

The main A.A. gun was the twin 40 mm Bofors. Known as the "Hazemeyer" this advanced mounting was tri-axially stabilised in order that a target could be kept in the sights on the pitching deck of a destroyer and was fitted with an analogue fire control computer and Radar Type 282, a metric range-finding set. It was a range-only type, so the mount could not blind fire.

The weapon was mounted amidships, on the position formerly used for a searchlight. This position gave the gun a good sky arc, but also exposed its delicate mechanism to spray. Maintenance was a nightmare. Standard outfit for ammunition stowage was 1,564 rounds per gun
In Royal Navy service the Hazemeyer was designated Mark IV. The Mark V was the 'utility twin' mounting.

Bofors - Hazemeyer

When, in May 1940, minelayer HNLMS *Willem van der Zaan* arrived in Great Britain after the fall of the Netherlands, her 40 mm close-range armament was inspected by several British representatives. The Hazemeyer F/C instruments were carried on the mounting. The guns, optical instruments and their operators were stabilised by gyro control of a triaxial mounting, control of fire being fully tachymetric. *"The officers who witnessed a demonstration were so impressed that one and finally two mountings were loaned for further tests on a rolling platform and manufacturing drawings to be made"*. Impressive!
It ought to have been a shock to men, who only a few months before had pronounced *"shooting weapons with sights … are adequate"*, to know that such accurately sighted A.A. guns were adopted by the Dutch before the war. If available in Holland, why not to Britain?
Source: British Destroyers by Edgar J. March, Seeley Service & Co. Ltd. 1966

The Mark IV was a self-contained twin mounting that had its own rangefinder, radar and analogue computer on the mount. This mounting used water-cooled guns and utilized a track and pinion system for elevating and training which was powered via a Ward-Leonard system for automatic target tracking. It was probably too advanced for its day and proved to be somewhat delicate for use. According to service notes, the Mk IV was apparently used more often in manual mode than in power mode. Maximum elevating speed was 25 degrees per second with the manually controlled joystick, but training and elevation via automatic control was limited to little more than 10 degrees per second.

Technical Data	
	40 mm Hazemeyer
Designation	Mark IV
Date in service	1942
Calibre	40 mm
Rate of fire	Practical: 80 to 90 rounds per minute
Max range (45°)	9,830 m (10,750 yards)
Shell weight	2.21 kg (4.88 lbs.)
Train rate	25 degrees / second
Elevation	-15 / +90 degrees
Train	360 degrees
Total weight	7.05 tons

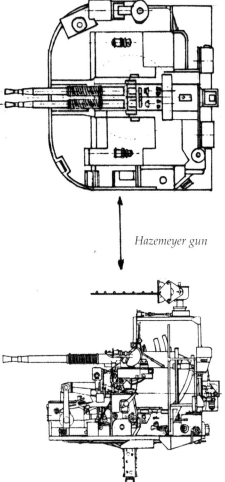

Hazemeyer gun

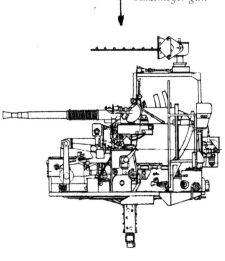

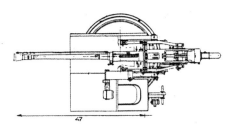

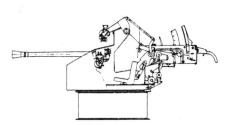

40 mm Bofors

40 mm Boffin

The Royal Navy also made extensive use of the Bofors. Its first examples were air-cooled versions quickly adapted for ships during the withdrawal from Norway. When a 40 mm Bofors gun was mounted on an Oerlikon power mount, it was called a Boffin. Boffins were mounted on each side of the forward superstructure and (after 1947) aft of the funnel. They were hand-operated, air-cooled single mountings.

The Boffin had a rate of fire of 120 rounds per minute and a maximum vertical height of 23,300 feet. Hydraulic pumps were mounted, which operated the 40 mm Boffins. Each gun had its own pump.

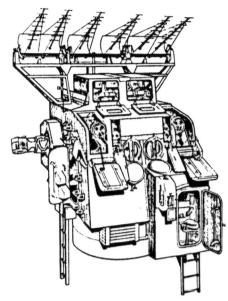

Forward superstructure of Kortenaer *while executing RAS in the late 1950s shows the portside platform and Boffin.*

FIRE CONTROL

20 mm Oerlikon

This was the close-range gun against aircraft and surface targets. The small-calibre guns had a moderate elevation the

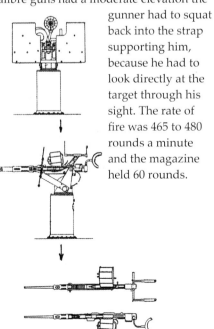

gunner had to squat back into the strap supporting him, because he had to look directly at the target through his sight. The rate of fire was 465 to 480 rounds a minute and the magazine held 60 rounds.

20 mm Oerlikon in fixed pedestal mounting

The ships of S-class retained the existing surface fire-control arrangements of earlier classes. There was a Destroyer DCT (Director Control Tower) on the bridge containing a gyro-stabilized sight and seats for a crew of 5 or 6. The type 285 DCT transmitted data for main armament

to a mechanical computer known as the Admiralty Fire Control Clock, which by a sequence of speciality designed cams and similar devices calculated the correct elevation and training for the guns. A bank of electrical transmitters conveyed these angles to the elevation and training receivers at the guns.

Type 285 was operating in UHF-Band and employed 6 Yagi antennas; three antennas ('fishbone') were used to transmit and three to receive. The bearing measurement was accurate but the tracking of aerial targets in elevation was very poor.

OTHER WEAPONS

Anti-submarine weapons

There was little space available on the quarterdeck for anti-submarine duties. Two overstern racks for depth charges and four depth charge throwers with adjacent reload stowages.

Torpedo Armament

The offensive power of the ships was by two quadruple torpedo mountings. British destroyers carried no reloads but had a small workshop on the upper deck where the delicate torpedo gyros where stowed; and limited maintenance on the 'fish' could be carried out on board. With the rear door of a tube hinged down and the torpedo partially withdrawn to the rear, some basic maintenance could be carried out. But the engine could not be test-run because it relied on surrounding sea water to keep it cool.

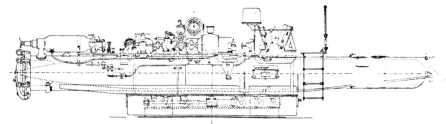

In 1956-57 Piet Hein *was reconstructed to a frigate with a helicopter platform midships but no hangar.* (NIMH)

Helicopter platform

The use of the heli-platform was a rarity, as the ship was not the most stable place to land and there were no facilities for service or maintenance.

The Sikorsky S-55 was transport helicopter for allround duties. Most of the time the carrier *Karel Doorman* had 2 (of the 3) embarked.

Photographs of helicopter landings on S-class frigates are rare.
On this page a Sikorski S-55 / H-19, the first real rescue helicopter of the Netherlands Fleet Arm is landing on Piet Hein *near Dutch New Guinea.*

The helicopter was one of the three in service. Delivered in 1954, named Cleopatra *(the others:* Salome *and* Delilah*) and stationed on carrier* Karel Doorman.
Cleopatra *crashed in February 1962 and had to be written off.*
(Collection: W. Vredeling)

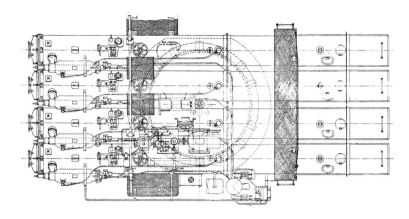

The tubes were always fired at 90° to the ship's centre line, so there was no need to aim precisely on a target. Normal control was from the Torpedo Sights on each side of the bridge. While launching the ship usually turned away in an arc from the target so that there was a spread in their tracks.

The forward quadruple had a control cupola for local control. *Serapis / Piet Hein* retained her tubes for all of her serving live.

Right:
Piet Hein launches a torpedo.
(F.W. Gerding)

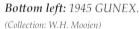

Bottom left: 1945 GUNEX.
(Collection: W.H. Moojen)

A break in the crews mess.
(Collection: W.H. Moojen)

Den Helder 1945. Piet Hein *arriving for the first time in Den Helder while a crowd is welcoming the navy's newest asset. (Collection: W.H. Moojen)*

Left: *In the nineteen fifties the navy invited a new generation to enlist with these kinds of ads in newspapers and magazines.*

In 1960 the Netherlands government hired tanker Mijdrecht *to provide the Tak Group 5 with oil at sea during their journey to the Far East. After arrival in Netherlands New guinea,* Piet Hein *was moored alongside. (Collection: W.H. Moojen)*

HMS Serapis / HNLMS Piet Hein

11 October 1943 - 12 June 1962

Commissioned	Named *Serapis*	First foreign port	Last foreign port call
11 October 1943	5th Ship	Kola Bay	Ouessant (7 June 1961)
Commissioned *Piet Hein*	**Named** *Piet Hein*	**Most distant port**	**Fate**
5 October 1945	8th Ship	Hong Kong (9,249 km)	Scrapped

1960s: War council. (Collection W.H. Moojen)

1958: In West New Guinea

Commanding Officers		
From	To	
1943	1944	Lt.Cdr. E. Lister Jones, DSC, RN
1944	1944	Capt. P.G. Lyon Cazalet, DSC, RN
1944	1945	Lt.Cdr. E. Lister Jones, DSC, RN
1945	1946	LTZ1 N.W. Sluijter.
1946	1947	KLTZ A. Molenaar
1947	1947	LTZ1 J.F.A. Pesie
1947	1948	LTZ1 Mzn. W. Vader
1948	1948	LTZ1 B. Hessing
1950	1951	LTZ1 A.P. Smitt
1951	1951	LTZ1 C.A. de Neef
1951	1952	KLTZ A.H.W. Von Freytag Drabbe
1952	1952	LTZ1 H. de Jonge van Ellemeet
1952	1952	KLTZ A.H.W. Von Freytag Drabbe
1952	1952	LTZ1 H. de Jonge van Ellemeet
1952	1953	KLTZ A.H.W. Von Freytag Drabbe
1954	1955	KLTZ A.M. Baron de Vos van Steenwijk
1955	1956	LTZ1 W.P. Coolhaas
1956	1956	LTZ1 A. v.d. Moer
1958	1959	LTZ1 J. Suermondt
1959	1960	KLTZ J.J. van Delden
1960	1961	KLTZ J.J. Kooijman
1961	1961	LTZ1 J, Burlage (temp.)

DOE WEL EN ZIE NIET OM

The badge is an allusion to the name of the Egyptian god, combining attributes of Osiris and Apis and having a widespread cult in Ptolemaic Egypt and ancient Greece.

The badge of Piet Hein *was derived from the crest of Piet Heyn, Admiral of the Maas Admiralty. Tradition has it that the arms are canting arms derived from his name: a bird (in Dutch: Piet) and an enclosure (Heyn).*

OPERATIONAL HISTORY

Ordered from Scotts & Co, of Greenock*
with 5th Emergency Flotilla on 9th January
1941. On 14 August 1941 the ship was laid
down, but not launched until 25 March
1943 due to serious design problems and
the late delivery of armament and fire-
control equipment.

SERAPIS

Following a successful Warship Week
National Savings campaign in March 1942
she was adopted by the civil community of
Prescot, Lancashire. Build was completed
on 23 December 1943.
After commissioning and workup, *Serapis*
joined the 23rd Destroyer Flotilla of the
Home Fleet based at Scapa Flow deployed
with screening duties in NW Approaches.

1944

On 22 February 1944, *Serapis* joined the
ocean escort for the Arctic Convoy JW 57,
(cruiser *Black Prince*, destroyers *Mahratta*,
Offa, *Matchless*, *Meteor*, *Milne*, *Obedient*,
Onslaught, *Oribi*, *Savage*, *Swift*, *Vigilant* and
Verulam). One escort and two U-boats were
sunk in the battles surrounding them. The

---

* Scotts Shipbuilding and Engineering Company
Limited, often referred to simply as Scotts,
was a Scottish shipbuilding company based in
Greenock on the River Clyde.

*Serapis anchoring
1944.*

*Artic conditions
for an unknown
emergency destroyer.*

In total, 104 Allied merchant ships were
sunk with the arctic convoys, along with
18 warships; 829 merchant mariners and
1,944 navy personnel were killed aboard
them.

42 merchantmen all reached Kola on 28
February.
1 March *Serapis* joined return Convoy
RA57. Arriving on 8 March and resumed
Home Fleet duties in NW Approaches
with Flotilla.

On 29 March joined escort for Russian
Convoy JW58 (with cruiser *Diadem*,
escort aircraft carriers *Activity* and *Tracker*
and destroyers *Obedient*, *Offa*, *Onslow*,
Opportune, *Oribi*, *Orwell*, *Saumarez*,
Scorpion, *Venus,* and Norwegian destroyer
Stord). On the 4 March, to the northwest of
Norway, the escorts damaged *U-472* which
was finished off by destroyer *Onslaught*. In

the next two days, in spite of foul weather,
they destroyed *U-366* and *U-973*.

On 7 April *Serapis* joined the escort for
return Convoy RA58 with the same ships
as JW58. Before arrival at Loch Ewe she
detached from the convoy on 13 April.
In May she was transferred with ships
of 23rd Destroyer Flotilla for support of
the landings in Normandy (Operation
Neptune) to be deployed in Force S.
Allocated target in Pre-Arranged Plan,
Lion-sur-Mer. *Serapis* took part in preliminary exercises with Force S ships.

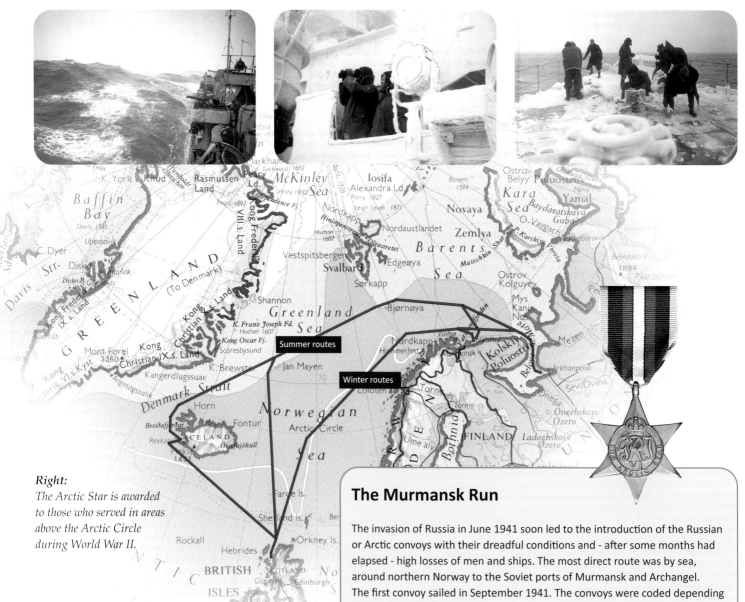

Right:
The Arctic Star is awarded
to those who served in areas
above the Arctic Circle
during World War II.

The Murmansk Run

The invasion of Russia in June 1941 soon led to the introduction of the Russian or Arctic convoys with their dreadful conditions and - after some months had elapsed - high losses of men and ships. The most direct route was by sea, around northern Norway to the Soviet ports of Murmansk and Archangel. The first convoy sailed in September 1941. The convoys were coded depending on their route initially **PQ** for outbound and **QP** for homebound. From 1943, the codes changed to **JW** and **RA**.

There were 78 convoys between August 1941 and May 1945, following the route through a narrow funnel between the Arctic ice pack and German bases in Norway. It was very dangerous, especially in winter when the ice moved further south. Many of the convoys were attacked by German submarines, aircraft, and warships.
Conditions were among the worst faced by any sailors. They were confronted with extreme cold, gales, and pack ice. The loss rate for ships was higher than any other Allied convoy route.

Over four million tons of supplies were delivered to the Russians. Although the supplies were valuable, the most important contribution made by the Arctic convoys was political. They proved that the Allies were committed to helping the Soviet Union, whilst deflecting Stalin's demands for a 'Second Front' (Allied invasion of Western Europe) until they were ready. The convoys also tied up a large part of Germany's dwindling naval and air forces.

Serapis escorted Artic convoys				
Departure	Joined	Convoy	Left Convoy	Arrival
20/02/44	22/02/44	JW 057	28/02/44	28/02/44
02/03/44	02/03/44	RA 057	08/03/44	10/03/44
27/03/44	29/03/44	JW 058	04/04/44	04/04/44
07/04/44	07/04/44	RA 058	13/04/44	14/04/44
31/10/44	02/11/44	JW 061A	06/11/44	06/11/44
11/11/44	11/11/44	RA 061A	17/11/44	17/11/44
30/12/44	01/01/45	JW 063	08/01/45	08/01/45
11/01/45	11/01/45	RA 063	18/01/45	21/01/45
03/02/45	06/02/45	JW 064	15/02/45	15/02/45
17/02/45	17/02/45	RA 064	27/02/45	28/02/45

On 3 June, deployed in the Solent with Force S. Departed on 5 June as an escort for convoy S2 comprising the 15th Mine-sweeper Flotilla, four Danlayers, two minesweeping motor launches, and an HDML (Harbour Defence Motor Launch). Escorted with *Scourge* and five Motor Torpedo Boats. After arrival the next day, she took up a bombardment position and provided support. Remained in the Eastern Task Force area for gunfire support, patrol and defence of the assault area until released from Operation Neptune.

After Normandy duties ended, *Serapis* resumed Home Fleet duties with the Flotilla at Scapa Flow. Deployed for escort of major units.

Selected for Operation Tenable. Escorting minelayers during mine laying off Norway on 29 September. The force reached its operational area in accordance with the plan, but the undertaking then had to be abandoned because of the sheer adversity of the weather.

14 October: *Serapis* escorted cruiser *Euryalus*, Aircraft Carriers *Trumpeter* and *Fencer* with destroyers *Myngs*, *Volage*, *Scorpion*, *Algonquin* (RCN), and *Sioux* (RCN) as Force 9 for aircraft mine laying and shipping strikes off Norway (Operation Lycidas). The force sailed from Scapa Flow. The carrierborne aircraft laid mines off Åramsund and Sado as a result of a navigational error on October 15, and in the designated area during the following day.

Operation Neptune

The code name for the naval phase of the Allied Invasion of Normandy during World War II. In total over 4,000 ships and a further 3,000 light craft took part in what was the largest seaborne invasion of all time: Operation Overlord.

6 June 1944 Naval forces were responsible for escorting and landing over 132,000 ground troops on the beaches. They also carried out bombardments on German coastal defences before and during the landings and provided artillery support for the invading troops.

Deployed as an escort on 24 October for cruiser *Devonshire*, Aircraft Carriers *Campania*, and *Trumpeter* with destroyer *Saumarez* as Force 2 for aircraft mine laying off Norway.

2 - 17 November: Deployed as a screen for cruisers *Berwick* and *Campania*, with destroyers *Cambrian*, *Caprice*, *Cassandra*, *Saumarez* and *Scourge* as Ocean Escort of Convoy JW61A comprising SS *Empress of Australia* and SS *Scythia* during passage to Murmansk. These ships were carrying Russian nationals who had been captured in Normandy. For the first time since September 1939, no merchant ships were lost throughout the length and breadth of the North and South Atlantic, including the Arctic, in October 1944.
From 7–12 December 1944, *Serapis* took part in Operation Urbane, a minelaying and anti-shipping survey conducted by the carriers *Implacable*, *Premier* and *Trumpeter*. The operation was a diversion during the RA.62 convoy operation (6/9 December 1944).

On 14 December *Serapis* escorted *Devonshire* and *Trumpeter* with destroyers *Zealous*, *Zephyr*, *Sioux* (RCN), *Algonquin* (RCN) and *Savage* as Force 2 for aircraft mine laying off Norway (Operation Lacerate)
The carrierborne aircraft laid mines in the Ramsøyund and attacked a number of coastal targets. However, the planned attacks on the airfields at Hammerfest and Banak, as well as the laying of mines in the Skatestrommen, were cancelled because of adverse weather and the fact that German aircraft were shadowing the force, which was attacked on three nights by Junkers Ju 88 torpedo bombers. One of these was shot down.

1945
From 1 January to 8 January 1945, *Serapis* escorted Convoy JW 63 to the Kola Inlet. Joined *Diadem* and *Vindex* with destroyers *Myngs*, *Savage*, *Scourge*, *Zambesi*, *Zebra*, *Algonquin*, *Sioux*, and Norwegian *Stord* as an escort for Russian Convoy JW63. Detached from JW63 on arrival. Joined escort for return Convoy RA63 with same ships as JW63 on 11 January. Both convoys passed through a total of 65 ships that month without loss.

Departed on 6 February joining Russian Convoy JW64 with Cruiser *Bellona*, Aircraft Escort Aircraft Carriers *Nairana* and *Campania*, 20th Escort Group and destroyers *Onslaught*, *Onslow*, *Opportune*, *Orwell*, *Zambesi*, *Zealous*, *Zest* and *Sioux* (RCN) as an escort. Detached on 15 February on arrival at Kola Inlet. Although JW64 reached Kola Inlet safely with all 26 merchantmen, the arriving corvette *Denbigh Castle* was torpedoed by *U-992* and became a total loss. Two days later *Serapis* joined the return Convoy RA64 as an escort with the same ships as JW64. Just off Kola Inlet *U-425* was sunk by sloop *Lark* and corvette *Alnwick Castle*, but later that day *Lark* was damaged by *U-963* and also became a total loss. Corvette *Bluebell* was then torpedoed by *U-711* and blew up with only one man surviving. Of the 34 ships with the convoy, one returned, one went down to U-boats and on the 23rd, straggler *Henry Bacon* was sunk by Ju88 torpedo bombers, the last ship of the war by German aircraft. The rest of the convoy arrived at Loch Ewe on the 28th after a voyage made even more difficult by violent storms typical of northern waters.
27 February: Detached from the convoy after an exceptionally stormy passage during which the convoy had to be re-assembled.

A Liberty Ship in convoy RA64 sails through heavy seas in the Arctic Ocean

On 19 March *Serapis* escorted Aircraft Carriers *Searcher* and *Queen*, cruiser *Bellona* with destroyers *Onslow*, *Zest*, *Haida* (RCN) and *Iroquois* (RCN) as Force 1 for an aircraft mine lay and shipping strike off Norway (Operation Cupola). The carrierborne aircraft attacked shore targets and laid mines off Askevold, the latter sinking one ship and causing others to be diverted farther out to sea, where they became targets for motor torpedo boat attack.

From 31 May to 17 June 1945, *Serapis* was refitted at Immingham and then returned to the Home Fleet, carrying out occupation duties at Wilhelmshaven in July.
On completion carried out post-refit trials and resumed Home Fleet duties with the flotilla at Scapa Flow

Scorpion was in the Battle of the North Cape on 26 December 1943, as part of the Arctic campaign. The German battleship Scharnhorst, *on an operation to attack Arctic Convoys of war materiel from the Western Allies to the Soviet Union, was brought to battle and sunk by the Royal Navy. During the battle,* Scorpion *and the Norwegian destroyer* Stord *attacked* Scharnhorst *with torpedoes, scoring two hits on the starboard side. The battle was the last between big-gun capital ships in the war between Britain and Germany.*

PIET HEIN

In autumn 1945, the Royal Navy announced that three S-class destroyers had been transferred to the Netherlands Navy. One of these, HMS *Serapis* entered the port of Amsterdam on 4 October 1945. The following day the official transfer took place. At that moment, she was one of the largest destroyers the Royal Netherlands Navy ever had.

Of the 230 crew, about 70% were War Volunteers (OVV) who joined the navy to contain the violence in the Indies after the Japanese capitulation and to restore colonial order. While others just signed up to join the fight against Japan in the summer of 1945.

On 22 October *Piet Hein* departed from Amsterdam for the Dutch East Indies. Initially, she was scheduled to sail with *Kortenaer* (transferred on Devonport 1 October), but due to a delay in preparations the ships sailed separately. She made a record passage of only 15½ days. At that moment a welcome reinforcement for the small fleet to monitor the large sea areas and counter piracy, smuggling, and the export of stolen merchandise. Arrived on 7 November in Tanjung Priok.

1946
Piet Hein sustained some damage. Since the Naval Establishment in Surabaya was still in ruins, the naval command sent the ship to Australia for maintenance. However, in Australia an unpleasant attitude began to emerge among the population, because of the troubles in the East Indies. In Australia, it was felt that the European colonial powers had been dismissed as world powers and "a good neighbour is better than a distant friend".

Anti-Indonesian independence propaganda poster "Indonesia Now! Draft as a volunteer"

5 October 1945: Commissioning Piet Hein.

Indonesian Revolution

The Indonesian War of Independence - took place between Indonesia's declaration of independence in 1945 and the Netherlands' transfer of sovereignty over the Dutch East Indies to the Republic of Indonesia at the end of 1949. The four-year struggle involved sporadic but bloody armed conflict and two major international diplomatic interventions. Dutch military forces (and, for a while, the forces of the World War II allies) were able to control the major towns, cities, and industrial assets in the Republican heartlands on Java and Sumatra but could not control the countryside. By 1949, international pressure on the Netherlands, the United States threatening to cut off all economic aid for World War II rebuilding efforts to the Netherlands, and the partial military stalemate became such that the Netherlands transferred sovereignty over the Dutch East Indies to the Republic of the United States of Indonesia.

The revolution marked the end of the colonial administration of the Dutch East Indies, except for New Guinea. It also significantly changed ethnic castes as well as reducing the power of many of the local rulers.

Piet Hein therefore made a number of unsuccessful attempts to call in to Australian ports for repairs. However, the workers firmly refused to do any work on the ship. Finally, in July 1946 the seriously leaking ship returned to Surabaya where the facilities were more or less restored. The necessary repairs were carried out there. More than a month later, the ship had recovered sufficiently, after which the patrol duties were resumed with broad powers. On Madura, the TNI made

shipping unsafe from a former Japanese bunker. It was shot into ruins by *Piet Hein*. (18 Aug) Actions followed at Semarang, the Makassar Strait and on the coast of Aceh. In the latter, coastal batteries were destroyed by guns.

1947

In January *Piet Hein* gave gun support to the army operations at Palembang, which were followed by patrols and various actions through the archipelago. In the

Bali Strait, a coastal battery that kept firing at passing ships was defused. In June, positions on the east coast of Java came under fire that had posed a threat to patrol boats and passing merchant ships. Finally, the ship also took part in the landing operations at Pasir Putih on 21 July, where preliminary shelling of the coast was conducted to support the advancing marines towards Probolinggo (First Police Action). With the necessary fire support, the ship sailed for a while

just below the coast. The ship was then directed to Cilacap, where, together with *Tjerk Hiddes*, she shelled the coastal batteries on Nusa Kambangan. Both ships landed forces to blow up the coastal batteries at Karang Bolong and Cimiring.

At the end of August, she finally departed homeward from Tajung Priok. For this journey, she embarked naval personnel for repatriation. Returning to Den Helder on 26 September.

1948
In September *Piet Hein* acted for a while as escort for aircraft carrier *Karel Doorman*. In December she was despatched to the Navy School for technical education in Amsterdam.

1949
Piet Hein and 14 other Neth. warships were dispatched a major exercise of the Western Union. Commencing 30 June, with the salutes to the flagship of each nation. Then on 4 July a fleet of 98 (!) warships departed for 3 days of manoeuvres in the English Channel and Bay of Biscay. On 7 July the ships entered Portland for debriefing.

Maintenance in 1949. Note the damaged buildings in the background.

1950

On 17 August departing from Rotterdam for a 6-month journey around the world. Sailing to the West. Visiting Willemstad, Netherlands Antilles (4 - 15 Sep) and passing the Panama Canal to Netherlands New Guinea.

1951

Returning to Den Helder and receiving maintenance.

1952

On 18 January 1952, *Piet Hein* departed under very unfavourable weather conditions (2.9° C, heavily clouded with a strong north-westerly wind). Besides the crowd of beloved ones, a Royal Marines Band played the National anthem when unmooring. A solemn moment in the stormy wind. Just outside port, an incoming wave knocked a quartermaster overboard. Although immediately responding, the search had to be abandoned after an hour. Knowing that given the weather conditions, the unfortunate quartermaster did not stand a chance. Course and speed were reassumed but shortly thereafter another incoming wave wrecked the starboard motorboat, while also the whaler sustained serious damage. *Piet Hein* manoeuvred to lie with her bow in the waves until the storm passed. On 19 January orders received to return for repairs.

After three days *Piet Hein* silently departed again. Now in improving weather conditions but still in a violent sea with strong winds to rendezvous with French units north of Oran for exercises. Unfortunately, the French warships had to return, but on 25 January the French air force proved to be a tough opponent. Early in the afternoon while *Piet Hein* proceeded to Malta, a division of destroyers of the U.S. Sixth Fleet passed heading to Gibraltar. She arrived on 28 January in Marsaxlokk Bay moored alongside the oil jetty. After a few hours course was set for Tripoli which was reached during the day watch of 29 January. A language barrier had to be broken when the local Arab pilot hardly understood any English.
Leaving Tripoli on the morning of 31 January, she headed for Port Said. A course was set along the coast. Arriving on the day watch of 3 February. The passage

through the canal took only 8 hours, thanks to the Commander-In-Chief Mediterranean in Malta, who gave cooperation to allow the ship to pass the channel as quickly as possible despite the convoy system that was set up due to the tense situation in Egypt. Port of Djibouti was called on the day watch of 7 February. Unfortunately, the quick passage resulted in a missing connection with the mail to the disappointment of the crew. The bags of mail had been forwarded to the next port of call since *Piet Hein* departed Djibouti the following day for Colombo. Sailing a route north of Socotra reaching Colombo on 13 February and passing Keppel Harbour for Singapore.

After replenishing she headed for Manilla on 18 February. Sailing with 18 knots east of Anamba Island and north of Prince of Wales and North Danger Reef with a

Korean War

The war began on 25 June 1950 when North Korea invaded South Korea following clashes along the border and rebellions in South Korea. North Korea was supported by China and the Soviet Union while South Korea was supported by the United Nations, principally the United States. No major naval engagements took place in the Korean War because the UN fleet controlled the coastal area from the moment they arrived on station. The Dutch destroyers and frigates showed their versatility in naval blockade, gunfire support, anti-aircraft fire and checking 'junks'. Throughout, they were harassed by enemy batteries and moored or 'drifted' mines.

The fighting ended with an armistice on 27 July 1953. The agreement created the Korean Demilitarized Zone (DMZ) to separate North and South Korea, but no peace treaty was signed, and the two Koreas are technically still at war, engaged in a frozen conflict.

R. Neth Navy in UN mission Korea		
From	To	Ship
07 Jul 1950	18 Apr 1951	Destroyer Evertsen
18 Apr 1951	21 Jan 1952	Destroyer Van Galen
21 Jan 1952	18 Jan 1953	Destroyer Piet Hein
18 Jan 1953	05 Nov 1953	Frigate Johan Maurits van Nassau
05 Nov 1953	10 Sep 1954	Frigate Dubois
10 Sep 1954	24 Jan 1955	Frigate Van Zijl

The first four ships have been awarded the Presidential Unit Citation of the Republic of Korea as unit of the US 7th Fleet.

Evertsen and Van Galen twice awarded the Presidential Unit Citation of the Republic of Korea as unit of Task Force 95.

A total of 4,748 Neth. military (1,360 navy) personnel have served in Korea. 121 were killed and 4 were missing in action. 68 were awarded:
- 3 Military William's-Order
- 5 Bronze Lion
- 21 Bronze Cross
- 4 Cross of Merit
- 14 Silver Star (US)
- 21 Bronze Star V (US)

A memorial in Honour of the Netherlands troops who fought in Korean War was built in 1975 at Ucheon-ri. The citation reads: 'Valiant fighters who acted in the spirit of the Prince of Orange and were filled with loyalty, 768 of them fell or were wounded in the struggle against the red invaders. This monument is dedicated to their determined fight'.

The Presidential Unit Citation

high-swell depression coming in from the northeast. This caused heavy stamping and incoming waves compelled her to slow down to eleven miles, while the wind and waves increased. On the day watch of 20 February, speed had to be reduced even further to 7.5 miles (maintaining 3.5 miles). On the next day, speed could gradually be increased again to 18 miles. In the meantime, a telegraph was sent to the Dutch envoy in Manila to inform the delayed arrival. Spotting the island of Luzon on 23 February, she carefully manoeuvred around the many shipwrecks without any difficulties while entering Manila Bay. Departing on 26 February for Hong Kong, she sailed on the route north of Stewart Bank, east of the Lena Islands and south of Victoria Island. Anchoring in Tathong Channel for the night and in the morning of 28 February, she entered the port of Hong Kong. Mooring at the Naval Dockyard to relieve destroyer *Van Galen* (D 803).

Korean War

On 27 March *Piet Hein* sailed for the first time to the Yellow Sea to relieve the frigate HMS *Crane* (U 23), which had been hit by

Bottom:
Piet Hein and two other destroyers in Korean waters.

a grenade on the previous day. It was now prohibited to walk on the landside deck and smoking was only allowed inside the ship. The patrol started near the island of Chodo. During the day, she carefully stayed out of the so called 'bombing line', closing in on the coast under the cover of night. Flares were used to spot targets too small to appear on radar and easily be used for infiltrations. The first patrol went smoothly.

On the last day of March, the alarm sounded long and emphatically. One of the US aircraft reported an assault over the ultra shortwave radio. At the same time some of the crew spotted small water fountains. A small South Korean vessel, towing another, transferring troops from Sosin to Chodo was under attack. Since the enemy battery was hidden from the

destroyer's view by a hill, an aerial reconnaissance and attack observation aircraft was launched by the carrier. *Piet Hein* remained at anchor but was ready to slip anchor if necessary. Although an indirect bombardment is usually carried out with only two guns, the commander decided to deploy all four guns to facilitate easier attack observation.

For the *Piet Hein*, this first artillery action under tactical command of the UN was in a way an experiment. To be frank, the gunnery officer was rather sceptical about the outcome of the shelling, because the crews consisted mostly of young sailors, who, due to the unfavourable weather conditions after leaving Colombo, had had practically no opportunity to practise and moreover, had been trained according to the British method of attack observation,

Delivery of mail by helicopter. Piet Hein *is on Plane Guard station, with the whale boat swung out to enable quick response to ditched aircraft. Note the national ensign painted on top of the B-gun.*

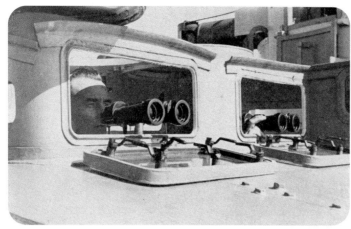

Gunnery control

while in Korean
waters the American method was used.
The "air spot" reported over the radio: -
'Sitting Duck, this is Duck shoot.
Gun-position near rocket hits. Over.
Two rockets fired by the aircraft indicated
the position of the target. At quarter to
ten, the first shells hurtled. After the
sixth salvo, the plane reported that the
target had been hit, just as the enemy was
ready to answer the fire. Five more salvos
followed, with corrections not exceeding
fifty yards. Fifteen minutes later, the target
had been destroyed, while two pillboxes
and three shelters were damaged. Cease
fire.

Meanwhile, *Piet Hein* was in the Task
Group with the US carrier *Bairoko*
(CVE-115) and participated in air operations on the west coast of Korea.
The third patrol was also to the West
Coast apparently still not occupied by the
enemy. Near Amgah, just north of the 38th
parallel, shells were fired at an enemy gun
emplacement. Also, the fourth and fifth
patrols were to the West Coast. Holding
post in the fighter screen of Task Group
95.1 with the carrier HMS *Ocean* (R 68)
during the day and patrolling along the
coast and between islands at night. Highly
depend on radar, because
at this time of year, there
was often a dense fog.

Yellow Sea

On the second patrol
to that sea area. A
powerful north-westerly
wind came up, bringing
with it clouds of dust
from the Gobi Desert,
whose bottom consists
of yellow loam. Within
hours, the entire ship
was covered with a
layer of this yellow dust.

9 July 1952, providing fire support to a US
minesweeper as it conducted sweeping
operations close to the enemy coast
before anchoring at Yeongdo. In the night
patrolling the coast in the area south of
Songjin, shelling a number of strategically important locations on the railway
line, which runs right along the coast here
and there and can be easily shelled from
the sea. During the day, the railway was
regularly bombed from the air by naval
and army aircraft, while naval forces had
to ensure that the enemy did not get the
chance to repair the damage at night. For
the supply of war material, this railway
was of inestimable importance to the communists, and it must be acknowledged that
they tried with great gusto to maintain the
connection with the South at all costs. If a
bridge was destroyed somewhere during
the day or a section of track was rendered
unusable by bombs, then after darkness
had fallen, many twinkling lights would
be evidence of the Reds' persistent efforts
to restart train traffic with all their might.
They usually succeeded in a few hours.
Nocturnal interference fire from the ships
was the only way to prevent this as much
as possible. The desire of every ship's
crew on the East Coast was to ambush
such a train and destroy it by gunfire. This
thought occupied everyone's minds all
the time, especially with the gun crews it
was a real "train psychosis". Occasionally
Piet Hein was ordered to open fire at shore
targets.

On the evening of July 15, 1952, *Piet Hein*
set out alone, to shell railway targets in
the area south of Songjin and if possible
close under the shore under cover of the
night darkness, ambush an enemy train
and destroy it with its artillery. There was
undeniably a sporting element to this kind
of patrol, the success of which depended
much on sharp observation and faultless
striking, while on the other hand the
darkness of the night only increased the
possibility of surprise, without giving any
certainty that an enemy coastal battery
would not open fire unexpectedly at such
a short distance. Everyone knew from
experience that there would probably be a
lot of gunnery that night and little sleep. A
railway tunnel was first in line.
19.45 h: The ship was darkened and
smoking on deck was prohibited.

20.20 h: Starboard watch located a target and guns 3 and 4 fired. As visibility was good, the bombardment was no more than a routine and soon completed, after which the patrol continued. The next target was a railway crossing, which provided an opportunity to open the hunt for trains. Indeed, on the previous night, the US destroyer had been lucky enough to neutralize a train laden with war materials. The success of the shelling lay primarily in stalling the enemy train connection with the battle front for as long as possible. This would not give the communists any opportunity to remove the wreckage and clear the railway. Therefore, a ship was posted in front of the site in question, with orders to thwart any attempts to carry out repair work through interference fire. That night, the destroyer USS *Endicott* (DD 495) was tasked, and the glow of flares could be seen from *Piet Hein*. There was a possibility that the enemy would try to maintain the broken supply line to the front by transferring the war material to road transport both in front of and behind the blocked road section. Such an opportunity to shoot yet another train off the rails was obviously not be missed!

21.30 h: Some distance inland, luminous phenomena were observed, which, in connection with the presence of a railway bridge there, suggested that an enemy repair crew was carrying out welding work. A short disruption fire of five salvos crackled. More suspicious lights appeared further along the railway, each of which was gassed on a few shells.

06.30 h: The next morning, *Piet Hein* took over the watch from *Endicott*. To give the impression that they had left the scene *Piet Hein* slowly sailed away. After covering about two miles, the destroyer turned sharply and snuck back to the bombardment position at high speed, simultaneously opening fire. Hordes of communist railroad men ran in panic along the runway, 'as the shells burst around them.

12.45 h: Ceased fire and set course for Yeongdo, leaving the enemy in doubt as to whether she was actually leaving

South to the Yellow Sea on 26 August 1952 Everything seemed to indicate a quiet patrol, with *Piet Hein*, steering varying courses in the Task Group east of Macau Island. On the afternoon of the 29th, the ship separated for a night patrol in the waters between Paengyong Do and Mahap To. However, that night she, received orders to sail to the island of Chamin To, pending further instructions. Fire support had to be provided for an amphibious operation on the enemy occupied Hoan Hui Do peninsula.

The attack was to target the command post of a Northern battalion stationed there, with the task of capturing military documents and bringing back as many prisoners as possible for intelligence. Initially, the landing was to be carried out at 05.00 h, but due to unforeseen circumstances, the troops went ashore 40 minutes later.

Left:
Storm in the Yellow Sea.

06.30 h: Battle stations sounded, and all hands came on post.
07.00 h: HMS *Ocean* launched its aircraft.

07.25 h: *Piet Hein* opens fire. Although the aircraft helped to identify the target, it was difficult to discern in the early morning mists, while the distance was a little less than the maximum firing range of the guns. Nevertheless, within two minutes, the first salvo was fired. The artillery distance was accurately determined to be 13,700 yards, to neutralise the enemy mortar positions.
07.40 h: HMNZS *Taupo* (F 423) reported 'Right on target!'
07.45 h: After firing 65 shells, the shelling stopped. Meanwhile, the aircraft had launched another attack
08.10 h: Radio message: Attack completed.
10.00 h: Course set to join Task Group.

In the afternoon of 14 November 1952, *Piet Hein* had replenished somewhere on the East Coast of Korea alongside USS *Passumpsic* (T-AO-107),

before sailing to the assigned location for a package sweep (a section of track located between two railway tunnels, which, due to its local conditions, has the potential to surprise and destroy an enemy train with artillery fire at night). Initially, the weather conditions were not favourable for the task: it was hazy and despite the short distance, little could be seen of the coast.

18.10 h: Alarm sounded. Steering at 2,500 yards offshore to obtain better land reconnaissance. Taking the shortest distance at which the mainland could be approached due to the likely presence of enemy minefields up to 2,000 yards offshore. In the evening, the land stood out like a jagged, dark mountain mass against the afterglow of the western bilge, with only the separation of land and water clearly visible. Lookouts peered tensely into the rapidly growing darkness. Meanwhile, the Asdic team continued its monotonous but important work. Using the short transmission device, the surrounding sea area was systematically pinged for the presence of mines. The ship kept up and down at two-mile speeds, but the engine room had both boilers on standby and was ready for extreme power should the need arise.
19.50 h: A lookout spotted a plume of smoke rising straight upward. Within minutes *Piet Hein* opened fire, but the train had already disappeared into a tunnel. The Artillery Officer was firmly convinced they had hit something. Since the Commander

Train Busters Club

Train busting was the art of bombarding trains on rail lines travelling along the Korean coast. Gunners had to be fast and accurate to hit the trains between the tunnels where they hid between salvos. It became much like a snap shooting in an arcade shooting gallery. In mid-July 1952, H.E. Baker, a U.S. naval captain and the Task Force commander, organized a club unique to the Korean War, the Train Busters Club. Baker started the club after his ship, the USS *Orleck*, destroyed two North Korean supply trains in a two-week period. Captain Baker immediately named the *Orleck* as the champion of the club. However, since the war was conducted under the UN banner, membership of the club was made available to any UN allied ship that destroyed a North Korean train.

To be admitted to the Train Busters Club, a train's engine had to be destroyed. Once that happened, any damage inflicted on a train would be counted as a kill regardless of the fate of the engine. The rules were enforced to the point that all kills recorded were considered legitimate.

Busting a train required a combination of shooting range, good weather, skilled gunnery crews, and on at least one occasion, sending a party of men in a small boat close to the shore to spot a train for the gun crews on the ship. An all-volunteer crew from the destroyer, USS *James E. Kyes*, set out on this mission. The crew was selected on the basis of familiarity with weapons. The boat officer, radioman, coxswain, and engineer had side arms. The commander of the crew carried a carbine.

Trains destroyed in Korea 25 June 1950 until the armistice on 27 July 1935:	
Crusader (HMCS)	4
Endicott (USS) 3	3
Piet Hein (HNLMS)	2
Athabaskan (HMCS)	2
Charity (HMS)	1
Haida (HMCS)	1
Carmick (USS)	1
Orleck (USS)	1
Pierce (USS)	1
Porter (USS)	1
Jarvis (USS)	1
Boyd (USS)	1
Trathen (USS)	1
Eversole (USS)	1
Keyes (USS)	1
Maddox (USS)	1
Chandler (USS)	1
Mc. Coy Reynolds (USS)	1
Total	25

TRAIN BUSTERS CLUB
TASK FORCE 95
THIS CERTIFICATE PRESENTED TO
HMNS Piet Hein, DD 805
FOR HER CONTRIBUTION TO THE UNITED NATIONS CAUSE AGAINST COMMUNIST AGGRESSION IN KOREA BY DESTROYING ONE COMMUNIST TRAIN
IN RECOGNITION OF A JOB WELL DONE
COMMANDER TASK FORCE 95

and the lookouts confirmed, the result was booked as a "suspected hit".

20.45 h: Another plume of smoke was detected. Ostensibly, the shells fell in the right area but in the glow of the flares fired in the meantime, it was nowhere to be seen. Someone mentioned the train had reversed into a tunnel, which would explain why most of the carriages were already hidden from view, while the plume of smoke was still visible. Anyway, the target was gone.

23.20 h: Port lookout spots another flume and within seconds a salvo was fired. But in the first flare, the train was nowhere to be seen.

03.31 h: Train sighted for a few moments, but before fire could be opened, it too had disappeared into a tunnel or behind a hill ...

04.10 h: When the ship was sailing in the opposite direction, the lookout spotted a train. Just 15 seconds after the first report, fire was opened at a distance of 4,100 yards. But the only smoke, which could be observed, was from the jumping grenades. In the light of a flare, they saw a train standing still. Now salvo after salvo crackled from the firing nozzles and

screeching shells rushed towards their target. It was an impressive sight to see a railway carriage explode, while flames burst from the wreckage. Every shot was a hit! This time there was no escape: the communist military train, which would live on in history as "Piet Hein's Train".

06.30 h: Cruiser USS *Helena* (CA-75) opened fire with her secondary battery of 4-inch guns on what remained of the enemy train.

Cold weather in the Yellow Sea

1953

After the relief by frigate *Johan Maurits van Nassau*, *Piet Hein* departed Hong Kong on 15 January. She had sailed 45,000 nautical miles and fired 3,000 11.9 cm shells in service of the United Nations.

Following a recommendation of a Merchant KJCPL-line inspector the home voyage went to Hainan using the coastline gulfstreams of Indochina and the calmer seas of the areas. On 19 January *Piet Hein* entered Singapore, berthing alongside a Shell tanker for replenishment and proceeded to the 'man of war anchorage'. The next day, in the afternoon, the ship departed again and headed for Malacca Strait. On 23 January, she entered Ceylon for a three-day visit before sailing to Aden (31 Jan) and Suez in deteriorating weather conditions. On 3 February *Piet Hein* anchored in the Bay of Suez, on standby to join a north bound convoy. She passed Port Said in a fog, and accelerating to 15 knots headed for Rhodes, Izmir (11 Feb) and Piraeus (12-16 Feb). Passing the Adriatic Sea and visiting Split (18-23 Feb), she continued to La Valletta (Malta) for replenishment and returned to Den Helder on 4 March.

1954

When in the early 1950's troubles arose about the sovereignty of Dutch New Guinea between Indonesia and the Netherlands the navy was increasingly confronted with infiltrations. In response, the warships were detached to the area.

4 March 1953, mission accomplished.

14 June 1955. A fleet review with 22 warships on the IJ in Amsterdam.
Front row left to right: S-class destroyers "Kortenaer, Evertsen and ASW destroyers Noord Brabant *and* Holland. *On the right 3 PCE and 2 inshore minesweepers, Back row: frigate* De Zeeuw. *minesweeper* Onverdroten *and cruiser* De Ruyter.

After months of maintenance *Piet Hein* left Den Helder on 17 August heading south to act as Duty Ship in Dutch New Guinea. Besides the crew, a scientist had embarked to study the country for the Rijksmuseum of Natural History. Passing the Dover Strait, some technical issues arose in the Gulf of Biscay. The port engine was put out of service and a stop was made in Cadiz where the 'grease monkeys' worked in shifts to repair it. Nevertheless, the visit needed to be extended 3 days (21-26 August). After departure, *Piet Hein* entered Gibraltar the same day for a few hours of replenishment. She sailed through the Gibraltar Strait in the late evening, carefully manoeuvring to avoid collision with one of the many poorly illuminated fishing vessels. On 1 September she arrived at Port Said and joined the south bound convoy next day to enter Suez on 3 September. By now the temperatures increased and yellow sand dust started to show all over the ship. Called in at Aden (6 – 8 Sept.) for steam valve repairs and Colombo (13 Sept.) before heading to Singapore (19 – 25 Sept.). On 30 September the most westerly part of Dutch New Guinea territory was entered, and a rendezvous was made with *Van Kinsbergen*. In company of the ships, she headed for Sorong, arriving on 30 September.

In the early morning of 21 October, a gang of 42 infiltrators landed with two boats at Omba, near Etna Bay. On the evening of the next day, the gang landed in kampong Etna Bay. A shop was looted. On 23 October the gang arrived in Kiruru and reached Kupai the next day. In response, the commander of the *Piet Hein* was appointed commander of 'Operation Etna'. A detachment of marines was flown in, occupied kampong Etna Bay on 25 October, and awaited the arrival of the *Piet Hein* the following day.

1955. Station ship in the Fast.

On arrival, a landing detachment of the destroyer with marines and police pursued the 42 Indonesian infiltrators. On 30 October a second police boat, the *Maro*, arrived with 13 police officers. A few days later the police boat *Memberano* arrived and former army boat *Van Ghent* was deployed. However, the latter was experiencing problems with the propeller shaft. All of the infiltrators, except for the leader with 2 men, were killed or captured. On 24 December 'Operation Etna' was ended and *Piet Hein* towed the heavily damaged *Van Ghent* to Kaimana.

1955

After the relief of *Van Kinsbergen*, the ship made some patrols in the area. Ending in December, the local Gouverneur and some high-ranking guests embarked for some days while sailing to the western part of New Guinea.

1956

Leaving Sorong on 16 March bound for the Api Passage near Sarawak for rendezvous with *Kortenaer*, the relieving duty ship. As the tradition goes a first radio contact was made with exchanging a bible reference. Once visual contact had been made, both ships anchored for some hours to brief the new arrival.

On 22 March *Piet Hein* entered Singapore for replenishment. After passing the Malacca Strait she headed for south of Sri Lanka (Ceylon) and followed the Indian west coast to Bombay. Meanwhile, the lookouts were occupied spotting the many small fishing vessels without any recognition lights.

Top and left page: Piet Hein *in Etna Bay.*

After visiting Bombay (30 Mar - 2 Apr.) she made her way to Karachi on a seemingly hazy day. But after a while, the crew noted the absence of moisture and the yellow sand. A light sandstorm did blow a quantity of sand all over the ship. Visiting Karachi (4 - 8 Apr). Without any warning, on the night of 6 - 7 April, *Piet Hein* found

herself in a severe storm (wind force 10) for over an hour. No harm was done/ Only the tents were forced to take a serious battering, during which some did not survive. The next morning, the port side of the ship was covered with a layer of sand. Shortly after a rain shower helped to clear the decks. She departed the next day for Aden (12 – 15 Apr.) and Suez (18 Apr), calling at Port Said for replenishment (19 – 20 Apr), Haifa (20 – 23 Apr), and Malta (26 Apr). Her speed had to be reduced because of unfavourable winds en route to Tangiers (30 Apr). She arrived at Den Helder on 8 May.

In 1956-1957 *Piet Hein* was reconstructed at the Royal Shipyard at Den Helder. She received a helicopter platform and was now designated as fast frigate with pennant number F 804.

1958

Commissioned on 1 July as fast frigate, on 12 September *Piet Hein* departed for a new torn in Dutch New Guinea. She sailed to Gibraltar (16 Sep.), Palermo (19 22 Sep.), Port Said (25 Sep), Aden (30 Sep), Colombo (7-10 Oct.), Singapore (14 – 18 Oct.) and arrived in Biak on 25 October.

1959

Because of the Indonesian claim and forthcoming tensions, *Piet Hein* often alternated patrols with combined exercises. On 16 January, the ship rescued the crew of a ditched Firefly VP 68.

1960

On 5 June a detachment of marines boarded on *Piet Hein,* leaving the next day for Port Moresby. Carefully evading Indonesians territorial waters, the destroyer headed for Sedeh Strait, north-east of Kai and Aru islands and Coral Sea to Port Moresby for an unofficial visit (9-12 June). After departure a strong south-east Passat

wind caused pitching and rolling in such a way that speed had to be reduced. Some mine damage was received. On 16 June anchoring near naval base Manus; HMAS *Tarangau* (16 - 18 June). On 21 June *Piet Hein* arrived in Biak.

On 1 November 1960 ASW destroyer *Amsterdam* and frigates *Kortenaer* and *Piet Hein* were stationed in Dutch New Guinea. During the night of 9-10 November 1960, an armed infiltration took place in the Boeroe area. In response, the destroyers *Amsterdam* and *Piet Hein* with marines embarked were directed to the area. On arrival, two platoons of marines disembarked for in land operations. After processing the gathered intel *Piet Hein* opened fire on the enemy's position in the forest forcing the enemy to flee in anger, leaving their weapons and ammunition behind. When the marines reached the location, they concluded the target was missed by

just a few metres. (On 24 March the action ended with the capture of the latter.)

On 24 November *Amsterdam* and *Piet Hein* captured the smuggling motor schooner

Maintenance in West New Guinea.

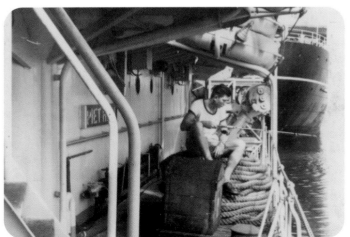

References

- British destroyers, *Edgar J. March, 1966, Seeley Service & Co. Ltd.*
- Conway's All the Worlds Fighting Ships 1922-1946, *Roger Chesneau (Editor), 1980 Conway Maritime Press*
- Destroyers of the Royal Navy 1893-1981, *Maurice Cockern 1981m Ian Allen Ltd.*
- *Lichtflitsen onder de kim,* C.J.M. Kretschmer de Wilde, 1955, Uitgevers-maatschappij W. de Haan nv.
- Jaarboek Koninklijke Marine *(various volumes)*
- Warship World *(various issues)*
- Website: *Dutchfleet.nl*

Series editor	Contributors
Jantinus Mulder	Dick Vries
Publisher	**Graphic design**
Walburg Pers/Lanasta	Jantinus Mulder
Authors	**Translation revision**
Jantinus Mulder	Deben Translations

First print, January 2024
ISBN 978-94-6456-192-0
e-ISBN 978-94-6456-193-7.
NUR 465

Contact Warship
jantinusmulder@walburgpers.nl

Lanasta

Sin Kuang. The crew of 5 was arrested and the ship brought up to Fak Fak.

1961

On 10 April, accompanied by destroyer *Amsterdam*, departing for the home voyage. On 13 April a rendezvous with *Evertsen*, on her way to New Guinea. Proceeding to Midway (19 - 21 Apr), Honolulu (24 - 24 (Apr), San Diego (4 - 8 May), Willemstad (20 - 26 May), Ponta Delgada (4 - 5 June) and Quessant. Arriving 10 June at Den Helder.

Stricken on 16 October 1961 and sold to Gebr. Van Heyghen for scrap in May 1962. She was broken up at Ghent (BE).